LET'S CODE!

CODING FUNCTIONS

BY GEORGE ANTHONY KULZ

CONTENT CONSULTANT
DAVID A. LASH
ASSOCIATE PROFESSOR OF COMPUTER SCIENCE
AURORA UNIVERSITY

Kids Core
An Imprint of Abdo Publishing
abdobooks.com

abdobooks.com

Published by Abdo Publishing, a division of ABDO, PO Box 398166, Minneapolis, Minnesota 55439. Copyright © 2024 by Abdo Consulting Group, Inc. International copyrights reserved in all countries. No part of this book may be reproduced in any form without written permission from the publisher. Kids Core™ is a trademark and logo of Abdo Publishing.

Printed in the United States of America, North Mankato, Minnesota.
102023
012024

THIS BOOK CONTAINS RECYCLED MATERIALS

Cover Photo: DC Studio/Shutterstock Images
Interior Photos: Shutterstock Images, 4–5, 12–13, 23, 28 (top), 28 (bottom), 29 (bottom); Faievych Vasyl/Shutterstock Images, 6; Tada Images/Shutterstock Images, 9; Chay Tee/Shutterstock Images, 10; Mike Flippo/Shutterstock Images, 14, 29 (top); Red Line Editorial, 17; Volodymyr Tverdokhlib/Shutterstock Images, 18; Jan Zwolinski/Shutterstock Images, 20–21; Miljan Zivkovic/Shutterstock Images, 24; Wavebreak Media/Shutterstock Images, 26

Editor: Katharine Hale
Series Designer: Katharine Hale

Library of Congress Control Number: 2023939667

Publisher's Cataloging-in-Publication Data

Names: Kulz, George Anthony, author.
Title: Coding functions / by George Anthony Kulz
Description: Minneapolis, Minnesota: Abdo Publishing, 2024 | Series: Let's code! | Includes online resources and index.
Identifiers: ISBN 9781098292768 (lib. bdg.) | ISBN 9798384910701 (ebook)
Subjects: LCSH: Coding theory--Juvenile literature. | Computer programming--Juvenile literature. | Computers and children--Juvenile literature. | Functions--Juvenile literature. | Analysis (Mathematics)--Juvenile literature.
Classification: DDC 005.1--dc23

CONTENTS

Food service workers are well-trained in following steps.

CHAPTER 1

STEP BY STEP

Talia works at a fast-food restaurant. Today, she is in charge of cheeseburgers. When a customer orders one, Talia's manager tells her to make a cheeseburger. Talia knows all the steps involved. First, she cooks a beef patty. She puts it on a bun.

Workers follow step-by-step instructions to make a cheeseburger. Functions are step-by-step instructions in computer code.

Then she adds cheese, lettuce, and tomato. She puts mustard and ketchup on top of that. Finally, she places another bun on top. She wraps it up and gives it to the customer.

Lots of people order cheeseburgers. Imagine if the manager had to tell Talia the steps to make one every time someone ordered. It would take a long time! Luckily, Talia already knows how to make a cheeseburger. The manager just needs to tell her to do it.

What Are Functions?

A similar idea is used in computer programming. Programmers are people who write code. Code is used in programs and **software**. These tell a computer what to do.

A set of steps to do a simple job in a program is called a function. Functions are written in code. "Make a cheeseburger" is like a function in Talia's job. She follows the steps to make a cheeseburger each time she is told.

Functions and Algorithms

Algorithms are also sets of instructions. Functions can be part of algorithms. Think of the process of ordering fast food. The customer orders a meal. One worker makes the cheeseburger. Another makes the fries. A third worker pours the drink. Then the food is given to the customer. The whole process is an algorithm. But each step is a function.

In a busy kitchen, each person has a job to do. Those jobs have instructions, just like functions.

Functions make it easier for programmers to write code.

Programmers write code to solve an overall problem. They use functions to solve a small part of that problem. Functions also let programmers repeat tasks without writing lots of code. Functions are not a coding **necessity**. Code will work without them. But functions are a helpful tool. They make writing code much easier.

Further Evidence

Look at the website below. Does it give any new evidence to support Chapter One?

Coding Activity: "Fashionable Functions"

abdocorelibrary.com/coding-functions

There are thousands of programming languages.

CHAPTER 2

PARTS OF A FUNCTION

Programmers have many ways to write code. There are different **programming languages**. These languages have different rules. But in most languages, functions look similar.

Functions have several parts. Every function has a name.

It is important for programmers to name their functions. The name lets the programmer call, or use, the function.

Programmers give their functions names when they write them. They use a function's name to call it. Calling a function tells the software to follow the function's steps. The software will leave the current line of code. It goes to the function and completes all the steps. Then it returns to where it left off.

Naming Rules

A function could be used to make a list. Programming languages have different **naming conventions**. Snake case is one example. In snake case, the function's name would be *make_list*. Another option is camel case. In lower camel case, the function's name would be *makeList*. In upper camel case or Pascal case, the function's name would be *MakeList*.

Many functions have inputs. Inputs are information that goes into the function. This may include numbers or words. The steps tell the function what to do with the inputs. The function could add numbers together. It could also put a list of words in alphabetical order. Functions also have outputs. The output is the result of the steps.

Working Together

Programmers can write functions for many tasks. For example, a programmer might write a function that teachers could use. The function finds a student's average grade. The student has grades for homework, tests, and projects.

Parts of a Function

Defining the Function

This section defines, or creates, the function.

Function name — Inputs

```
define subtract_numbers(num1, num2)
  difference = num1 - num2
  print ("Difference:", difference)
```

Steps

Output

Function in Action

```
subtract_numbers(8, 6)


Difference: 2
```

The function is called by typing its name. The inputs go inside the parentheses.

The output will print.

This sample function subtracts one number from another and outputs the difference.

Keeping track of a student's grade requires math. Functions can help.

These are the inputs. The steps of the function perform math to find the average. To do this, the function adds the grades together. Then it divides by the total number of grades. The output is the student's average overall grade.

Explore Online

Watch the video at the website below. Does it give new information about functions that wasn't in Chapter Two?

What Is a Function? Using Functions to Code

abdocorelibrary.com/coding-functions

When putting together a puzzle, it helps to focus on one part at a time. Then the small parts come together to make the big picture. Functions help programmers do this with their code.

CHAPTER 3

USING FUNCTIONS

There are many reasons why programmers use functions. One is problem-solving. Programmers usually have a big problem to solve with their code. Functions allow programmers to break the problem into smaller parts. They can solve one part at a time.

Functions are useful in creating video games.

Functions in Action

A programmer could be working on a game. The programmer needs a character to walk. They write code to make this happen. But the character walks a lot. The code would get too long if the programmer had to write the same instructions every time the character moved. Instead, they can write a "walk" function. That function can be called each time the character needs to walk. Another person might be programming a different character in the game. They can use the same "walk" function.

Another programmer might want their program to create a rectangle on screen. They write code to create the four lines of the rectangle. But maybe they want to make several rectangles. They don't want to write code for all four lines every time. Instead, they create a "rectangle" function. The function automatically draws a rectangle if it's told where the corners are.

There are many benefits to using functions in programming.

Functions are important for programmers. They are not strictly necessary. But they have many benefits. They help break a program down into smaller pieces. They help avoid repeating code. They also encourage code sharing. Using functions makes coding better.

PRIMARY SOURCE

Dipam Vasani is a computer programmer. In an article about writing better functions, Vasani says:

> The first point to always remember when writing a function is to make them do one thing and just one thing. This tip will help you keep your functions short and simple. Don't put everything in it.

Source: Dipam Vasani. "A Walkthrough for Writing Better Functions." *Towards Data Science*, 13 Mar. 2020, towardsdatascience.com. Accessed 17 Apr. 2023.

What's the Big Idea?

What is this quote's main idea? Explain how the main idea is supported by details.

CODING DATA

A function is a set of steps to do a simple job in a program.

Programmers use functions so they do not have to repeat lines of code.

Programmers give a function a name. They use the name to call the function.

Functions can be shared with other programmers.

Glossary

algorithm
a set of instructions used to complete a program or task

bugs
problems in computer code that keep it from working correctly

naming conventions
in programming, rules for how to write the names of different parts of code

necessity
something that is needed or necessary

programming language
a vocabulary and set of rules for writing computer programs

software
all of the programs that give a computer instructions

Online Resources

To learn more about functions, visit our free resource websites below.

Core Library CONNECTION
FREE! COMMON CORE MULTIMEDIA RESOURCES

Visit **abdocorelibrary.com** or scan this QR code for free Common Core resources for teachers and students, including vetted activities, multimedia, and booklinks, for deeper subject comprehension.

Booklinks NONFICTION NETWORK
FREE! ONLINE NONFICTION RESOURCES

Visit **abdobooklinks.com** or scan this QR code for free additional online weblinks for further learning. These links are routinely monitored and updated to provide the most current information available.

Learn More

Bell, Samantha S. *Coding Algorithms.* Abdo, 2024.

González, Echo Elise. *Organizing Data.* World Book, 2021.

Wainewright, Max. *25 Scratch 3 Games for Kids: A Playful Guide to Coding.* No Starch Press, 2019.

Index

About the Author

George Anthony Kulz holds a master's degree in computer engineering. He is a member of the Society of Children's Book Writers and Illustrators and has taken courses at the Institute of Children's Literature and the Gotham Writers' Workshop. He writes for children and adults.